The Architecture of Discipline:

Building Integrity in the Intelligent Cloud

Lori Higham

Author's Note

This work reflects lessons drawn from leading the design and governance of systems that must demonstrate their own integrity—daily, at scale, and under scrutiny. Architecture is not only a theoretical discipline but the structure through which accuracy becomes observable.

Across years in cloud security and enterprise governance, I've seen that technical excellence alone cannot preserve coherence. Only disciplined design, encoded verification, and a culture of accountability can.

The intent of *The Architecture of Discipline* is to restore architecture to its essential purpose: to make integrity observable. In an era defined by AI and Zero Trust, assurance replaces assumption; only systems that can demonstrate their integrity deserve to be trusted.

Acknowledgments

This work reflects the discipline and insight of many who have chosen structure over convenience and integrity over speed.

To my former Cloud Security team, whose passion for strategic architecture, ethical approach to cloud, and friendship I will always remember; to my Board of Advisors at Secure Cloud Provider, whose guidance and insights are invaluable; to my Encryption friends, whose commitment to security has always rivaled my own; to Samrat Patel, for his strategic insight and outstanding teamwork; to Vaughn Selvey, for his deep expertise in government standards and federal contract requirements; and to the many other brilliant leaders, architects, and engineers who practice verification as a craft—your rigor shaped the questions that became this book.

I am grateful to the peers and mentors who challenged assumptions, tested definitions, and reminded me that clarity is a form of respect.

To those advancing governance and security in the intelligent cloud, your work affirms that discipline remains the foundation of progress. You build systems that earn trust not through assertion but through transparency, consistency, and adaptation.

Finally, to those carrying these ideas forward: may your architectures not only endure but evolve—advancing as rapidly and responsibly as the technology they are intended to govern.

Contents

Chapter 1—Foundations and Role of Architecture

Cloud adoption is accelerating faster than most organizations can design for it. Workloads, data, and identities now move to cloud platforms at a velocity that exceeds the evolution of the architectures meant to contain them. Each migration expands capability, but in equal measure, exposure; and few enterprises have matured their design discipline quickly enough to manage that widening differential.

What presents as agility often conceals unmanaged complexity. New services, integrations, and layers of automation extend what the organization can accomplish, but simultaneously multiply what it must constrain. Every new dependency creates a junction of trust, and every junction another potential point of drift, cost, or failure. As the architecture's mass increases, governance begins to strain under the very mechanisms designed to sustain it.

When that strain becomes invisible, how would you know? When delivery accelerates beyond the rate at which structure can adapt, at what point does momentum begin to erode control?

Architecture is the mechanism through which enterprise intent is translated into enforceable and verifiable form. Without that translation loop, progress becomes the illusion of advancement detached from discipline. Governance, therefore, is not bureaucracy but design in motion: the capacity to define, implement, and sustain intent through systems that remain accountable to their own structure.

Architecture as a System of Discipline

When understood correctly, architecture functions not as a collection of diagrams or design preferences but as a dynamic system of control. It governs the transformation of intent into behavior and of behavior into verifiable outcome. Within this system, three mechanisms are inseparable: **definition, enforcement**, and **measurement.**

Definition establishes what "good" means: the principles and standards that express an organization's tolerance for risk and its interpretation of integrity.

Enforcement translates those expectations into configuration, automation, and policy, embedding control into operation rather than supervision.

Measurement supplies the continuous evidence that those controls remain effective, converting assurance from periodic declaration to ongoing validation.

Together they form the architecture of discipline: definition sets direction, enforcement applies it, and measurement sustains it. Remove any one, and governance collapses into assumption. Without definition, controls lack purpose; without enforcement, standards remain aspirational; without measurement, truth reverts to belief.

Reflect:

Where in your current environment is control presumed but unverified?

Can your architecture validate, in operation, that what you believe about your systems is still true?

Boundaries as the First Control Plane

The foundation of mature architecture lies not in tool choice or provider selection but in the precision of its boundaries. Each environment must exist within a declared scope of control: defined ownership, applied standards, and observable deviation. In distributed ecosystems, trust cannot be inferred; it must be engineered.

Architecture defines that engineered trust: where it begins and ends, how it is verified, and what evidence substantiates it. These boundaries are not constraints on autonomy but on the very conditions that make autonomy possible. Without them, even disciplined teams drift into contradiction, each optimizing for speed or local efficiency in ways that incrementally erode collective coherence.

When boundaries blur, small inconsistencies propagate. A policy differs slightly across accounts; an encryption standard is applied manually; a baseline check is deferred for expediency. None of these acts seem consequential in isolation, yet collectively they form the seams through which risk moves fastest. By the time visibility returns—often through an incident report rather than a telemetry feed—ownership disputes have already replaced prevention.

Architectural oversight exists to perceive what distributed teams cannot: the aggregate behavior of the system. From that vantage point, patterns reveal themselves: policy drift, conflicting automation, control gaps that span tool boundaries. Without that systemic perspective, governance degrades into a mosaic of local optimizations, each justified individually yet misaligned in total.

Reflect:

Are your cloud boundaries defined by design or discovered through incident?

Who within your organization possesses the vantage point that perceives the whole rather than the assembled fragments?

Complexity and the Cost of Drift

Loss of control seldom begins with a single failure. It begins with accumulation: a series of rational, localized decisions that compound until the system's shape no longer matches its design. A new tool fills an immediate gap; a configuration is altered to meet a deadline; an acquisition arrives with its own inherited stack. Each act feels pragmatic, yet together they alter the topology of governance itself.

At scale, inconsistency ceases to be inefficiency and becomes exposure. Deviations create decisions that can no longer be seen or validated. What cannot be validated cannot be governed, and what cannot be governed will eventually fail under pressure.

Architecture counteracts that entropy through shared operating models—standardized identity hierarchies, consistent segmentation, centralized logging, and inherited policy frameworks. These are the containment structures that translate diversity into coherence and coherence into reliability. Without them, energy diverts from innovation to reconciliation, as teams spend cycles aligning differences rather than advancing design. The irony is enduring: the pursuit of speed, unmanaged, becomes the drag that impedes it.

Reflect:
How many temporary exceptions have become permanent? How much of your architecture operates by standard, and how much by habit?

From Policy to Execution

Policies and strategies possess no power until design renders them executable. Architecture gives them substance—translating intent into constructs that can be built, tested, and verified. Through infrastructure-as-code and policy-as-code, governance ceases to rely on persuasion and becomes intrinsic to operation.

When design decisions are versioned, reviewed, and validated like software, governance transforms from aspiration into evidence. The architect's role evolves accordingly: not as recorder of desired states but as engineer of assurance—the conduit through which executive intent becomes operational behavior. Without that bridge, policy remains documentation; with it, discipline becomes self-sustaining.

Reflect:
How many of your standards enforce themselves by design?

How much of your governance still depends on documentation and trust?

Measurement as Continuous Verification

Measurement is too often mistaken for reporting, yet it is the dynamic pulse of control. Without evidence, assurance is speculation.

Mature organizations instrument their architectures to validate compliance continuously, collecting telemetry at enforcement points and surfacing deviation in real time. Metrics such as drift rate, remediation interval, and control coverage quantify architectural health, distinguishing activity from assurance and motion from stability.

Automated evidence generation turns audit from retrospective investigation into confirmation of what the system already knows to be true. Verification becomes continuous; inspection becomes formal acknowledgment.

Reflect:
Do your metrics demonstrate control or merely movement? Can you measure the health of your architecture with the same precision you apply to performance or cost?

Leadership and the Ethics of Control

When leadership perceives architecture as a control system rather than a technical domain, governance becomes an enterprise language—shared, empirical, and measurable. The conversation shifts from *Are we compliant?* to *How is compliance sustained and demonstrated?* One asks for state; the other for continuity.

At this level, architecture acquires ethical dimension. It is not only structural assurance but discipline in practice, the mechanism by which trust among teams, systems, and stakeholders is continuously renewed. Integrity, once embedded into design, ceases to rely on assertion; it becomes the natural outcome of disciplined behavior.

Reflect:
When assurance is requested, can your architecture produce validation of control or only declarations of intent?

Discipline as the Foundation of Integrity

The foundation of stability is discipline: the deliberate containment of complexity through repeatable, enforceable design. Discipline is not rigidity but predictability under change. It ensures that evolution occurs within defined bounds, preserving integrity as technology shifts beneath it.

Through discipline, governance becomes reflex rather than reaction, transforming enforcement into habit and validation into continuity. Architecture, in this form, ceases to merely document compliance and demonstrates it through consistent behavior. What begins as the governance of systems matures into governance through systems, with discipline embodied in design.

Closing Reflection:

What evidence demonstrates that your controls function as intended?

If those controls were to fail, how quickly would your architecture reveal it, and what form would that validation take?

Chapter 2—Standards and Governance Mechanisms

Governance often fails even when controls are present, because the underlying standards that serve as a basis for those controls are undefined, outdated, or unenforced. In other cases, organizations mistake documentation for control, assuming that intent will translate into design fidelity. It never does.

What differentiates effective governance from bureaucracy is enforceability: the ability to translate expectations into measurable, automated, and testable control. Architecture provides that mechanism of discipline, ensuring that intent does not remain aspirational but becomes reproducible in practice.

Standards as the Architectural Blueprint

Standards exist to preserve coherence so that change, however rapid, does not dismantle integrity. They are the operational specifications that transform principle into constraint, defining how quality, security, and interoperability are maintained across diverse systems and teams.

Frameworks provide alignment; standards provide precision. Frameworks articulate what must exist; standards determine how it is achieved. One establishes external accountability, the other internal repeatability. Together they form the joint through which autonomy and conformity stay in equilibrium, allowing innovation without the erosion of control.

In mature enterprises, standards are not static decrees but dynamic contracts between intent and implementation. When those contracts stagnate, governance fractures, and teams fill the void with local interpretation. Consistency degrades into divergence long before oversight detects it.

Reflect:
Are your standards engineered for enforcement or written for reference?
Do they evolve at the pace of the technologies they govern?

The Baseline Standard as Source of Truth

Every organization requires a single, authoritative source of configuration truth: its baseline standard. The baseline serves as the contract between enterprise and execution, defining encryption requirements, identity hierarchies, logging depth, and segmentation models.

Its strength lies in clarity. Ambiguity is the enemy of automation. A baseline that depends on interpretation ensures inconsistency, while one that is executable ensures discipline. When compliance relies on human judgment, deviation becomes inevitable.

The essential question is whether the baseline can be enforced by code or only interpreted by people. The former scales; the latter erodes. This distinction defines the boundary between aspiration and control, the threshold where governance becomes operational reality.

Reflect:
How many exceptions persist because standards invite subjectivity rather than precision?

The Cost of Incomplete Enforcement

Incomplete enforcement creates the illusion of control. Policies may exist, yet when coverage is partial or uneven, risk enters through configuration gaps and interpretive differences. Drift begins not through neglect but through entropy, the gradual divergence between design and deployment.

The cost is measured not only in incidents but in confidence. Architecture loses its predictive power, and decisions are made on assumptions no longer grounded in system reality.

To sustain integrity, execution must be centralized, traceable, and measurable. Without that structure, each team implements its own version of discipline, and governance regresses from prevention to post-event verification.

Reflect:
Can your organization demonstrate what was enforced, or only what was intended?
Is baseline enforcement automated, or dependent on diligence and goodwill?

Traceability: The Architecture of Assurance

Standards gain meaning only within a traceable system of accountability. When that lineage breaks, governance becomes paperwork: frameworks detached from policy, policy detached from control, control detached from evidence.

A coherent architecture maintains visible inheritance: framework, standard, control, and validation.
This continuous chain produces both defensibility and insight. Defensibility demonstrates compliance externally; insight reveals control performance internally.

When traceability weakens, enforcement may continue, but governance loses context. Activity replaces assurance, and measurement loses interpretive value. Traceability is therefore not bureaucracy but geometry, the alignment that gives the system its shape and its capacity to be understood.

Reflect:
Can every enforced configuration be traced to a governing requirement?
Where does that lineage reside, and who ensures its continuity?

Continuous Validation and Architectural Feedback

Even the most rigorous standards degrade over time. Platforms evolve, and assumptions once embedded in policy shift away from the realities of implementation. Continuous validation tools, such as Azure Defender for Cloud, AWS Config, Security

Hub, and GCP Security Command Center, reveal these changes by exposing whether controls remain aligned with their design intent.

Telemetry, however, should serve as more than confirmation. Used correctly, it becomes architectural feedback: data that informs refinement, identifying where standards lag and assumptions no longer hold.

This is where discipline matures into resilience. Detection precedes incident; adaptation replaces explanation. The faster deviation is identified and reconciled, the more continuity governance attains.

Reflect:
Does your telemetry merely verify compliance, or does it guide architectural evolution?

Governance as a Dynamic Lifecycle

Governance is not a project but a continuous loop that sustains alignment between architecture and reality. Each stage sustains the next:

1. Define standards from risk and obligation so expectations are explicit.

2. Implement them through automated enforcement so intent becomes behavior.

3. Measure adherence continuously so assurance is empirical, not assumed.

4. Refine standards as environments evolve so discipline remains contemporaneous with change.

When these stages connect without interruption, evidence displaces assumption and governance becomes a process of renewal. Enforcement generates data; data reveals deviation; deviation informs design.

In practice, few organizations sustain the loop end-to-end. Many define without automation, collect without analysis, measure without synthesis. The result is latency: discipline delayed until exposure emerges.

Reflect:
Does your governance operate as a feedback loop or a recurring reset?
How many of your controls have improved through insight rather than incident?

Architecture as the Engine of Governance

Governance mechanisms are the operational instruments of architecture, the machinery that converts principle into predictable behavior. When implemented well, deployments validate configuration against approved baselines before execution, flagging deviation for immediate review.

As delivery accelerates, architecture must deliver discipline at equal velocity. Its success depends not on vigilance but on structure embedded within workflow: policy-as-code, automated attestation, and continuous verification.

In this state, governance becomes reflex, design enforcing itself through motion rather than mandate.

Closing Reflection:
Can your controls sustain governance at the same velocity as change?
Where automation accelerates delivery, does discipline accelerate with it, or fragment beneath it?

Chapter 3—Automation, Evidence, and Continuous Governance

Automation is the bridge between policy and practice, translating defined standards into consistent, repeatable execution across pipelines and infrastructure. Without it, even the best governance models collapse under the strain of human interpretation and manual drift.

Manual enforcement cannot keep pace with modern delivery. As deployments multiply across accounts and regions, checklists and approvals fall behind almost instantly. The faster the pipeline, the smaller the margin for oversight, and the greater the risk of undetected deviation.

Architecture reveals a fundamental principle: a control that cannot be automated cannot be guaranteed. Automation shifts compliance from an event to a continuous behavior, embedding checks within the same mechanisms that deliver change. Yet, importantly, automation alone does not create control; pipelines without controlled configuration merely reproduce mistakes at scale.

Reflect:
Is your automation enforcing standards, or accelerating misconfigurations?

The Role of Automation in Governance

Automation makes governance enforceable by replacing interpretation with verification and transforming compliance from a best effort into an expected outcome. Once automation begins to enforce what teams once managed manually, the rhythm of delivery inevitably changes. Deployments may initially fail where standards are now applied without exception.

In practice, automation can slow delivery before it stabilizes it. Pipelines must be rewritten, controls refactored, and exceptions redefined. That friction is necessary; it exposes weaknesses that manual processes conceal. It also demands rigorous validation before production deployment because once a permissive rule or flawed configuration is codified, it scales faster than review processes can respond.

Without feedback, automation creates only the illusion of control. Reliable automation depends on continuous verification against a clearly defined, strategically designed baseline. When done properly, this cadence surfaces risk early and often corrects it, transforming automation from a mechanism of speed into an instrument of discipline.

Reflect:
Are your automation systems structured so that errors are detected and contained before they can propagate into production?

Measurement and Evidence of Control

Architectural metrics reveal how design performs under stress, tracing outcomes back to strategic decisions. The more difficult question is what those numbers truly represent. Metrics may suggest stability while reflecting only a limited scope of observation. Even when indicators appear favorable, a steady compliance rate can mask the growth of exceptions rather than signal improvement.

Evidence of control is relational: detection must lead to action, action to validation, and validation back to design. Without that chain, measurement records activity but contributes nothing to architectural learning or resilience.

Metrics therefore serve as the sensory system of governance. They show not what exists, but how faithfully behavior aligns with design intent.

Reflect:
Where controls appear stable, what evidence confirms that they are keeping pace with architectural change?

Continuous Governance as an Operating Model

Continuous governance integrates automation and measurement into a responsive control structure. Cloud-native policy frameworks—AWS Service Control Policies, Azure Policy, and GCP Organization Policies—enforce both preventive and corrective actions. High-risk operations can be blocked outright, while conditional policies trigger remediation when defined events occur.

Traditional governance detects deviation only after it has propagated; continuous governance compresses that delay by embedding detection and response directly within the architecture. Configuration scanning, posture management, and telemetry streams supply the evidence; policy engines interpret that evidence and act upon it.

The result is dynamic risk management: earlier visibility, faster remediation, and continuous validation of state. Yet as automation assumes greater authority, the trust boundary between human and machine decision-making becomes more complex and more critical to govern.

Reflect:
When a control executes automatically, what ensures that its outcome still aligns with design intent?

The Human Element of Automation

Automation expands the reach of control, but it also transforms the human role within that system. As enforcement shifts into code, human responsibility evolves from execution to verification. People ensure that automation operates within design intent and continues to represent current risk and policy.

Human expertise defines the boundaries of automation, determining where policies and remediation scripts must be reviewed, tested, and adjusted as systems evolve. When that oversight weakens, automation may continue enforcing conditions that no longer reflect reality.

Architecture benefits from an automation control plane that unites telemetry with human review. Tracking failed policy applications, rollback frequency, and exception handling provides quantitative evidence of performance; human interpretation converts that evidence into assurance, validating that enforcement remains both correct and relevant.

Reflect:
How does your organization detect and respond when automation diverges from expected behavior?

Automation as an Expression of Architecture

Automation has become the language through which architecture operates. It no longer merely enforces design; it expresses design, rendered in executable form. Cloud policies, infrastructure-as-code templates, and pipeline logic define how principles translate into behavior and how quickly architecture adapts to change.

This convergence shifts the role of architecture from documentation to orchestration. Standards are not only written and referenced but also versioned, executed, and continuously verified.

However, automation evolves faster than the intent it represents. Without a clear link between architectural principle and automated enforcement, systems begin to exhibit patterns that no longer align with organizational priorities.

To maintain coherence, architecture must establish checkpoints where telemetry and human review confirm that automation still embodies intent and reflects present risk.

Reflect:
When automation performs flawlessly, how do you verify that it is still right?

Closing Observation

Automation sustains governance by converting principle into persistence. Yet it also amplifies error when left without oversight. Continuous verification—technical and human—is what ensures that discipline does not decay as speed increases.

When automation operates within defined standards, it preserves coherence. When it acts without structure, it generates the very disorder it was designed to contain.

Chapter 4—Managing Complexity: The Architecture of Rationalization

Even the most disciplined automation cannot withstand unmanaged complexity. Complexity rarely arrives as chaos; it accumulates in increments that each appear rational. A new tool solves a local problem, a configuration extends to meet a deadline, a connector satisfies a specific use case. None of these choices seems significant alone, yet when combined and left unbounded by architectural constraint, they begin to distort the system's form. What once resembled design becomes accretion: layers of functionality whose interactions no one can fully explain and whose governance no single owner can enforce. Complexity is not the consequence of growth; it is growth without discipline.

Every tool introduced into an environment carries a cost that exceeds its license. It expands authentication surfaces, introduces new tokens and data paths, and extends the perimeter of trust in ways that are rarely visible. Over time, the organization realizes it has multiplied dependency faster than it has strengthened assurance. Oversight thins, and what was meant to increase capability instead multiplies fragility.

Architecture exists to impose limits upon that slope. It defines, with precision, where capability ends and risk begins. Every decision to implement, retain, or retire a tool must therefore be architectural—because each alters the system's topology, evidence path, and capacity for control. When such decisions are distributed without coordination, the enterprise no longer possesses a single architecture but a federation of assumptions, each optimized for its own version of "good enough."

Reflect:
When every addition promises acceleration, who decides when the system must slow down to remain safe?

The Cost of Unmanaged Complexity

The first symptom of unmanaged complexity is not failure but friction. Teams begin to duplicate effort without realizing it. The same function—compliance scanning, identity provisioning, or telemetry aggregation—is performed by multiple tools, each with partial adoption and conflicting data. The illusion of visibility multiplies while true observability declines.

Operationally, this drag converts engineering time into reconciliation time. Financially, it disguises redundancy as investment.

Redundancy matures into risk. A misconfiguration detected by one tool but ignored by another creates uncertainty—was the deviation real or merely out of scope? In that ambiguity, assurance weakens. Responsibility disperses across teams that act independently, each guided by its own telemetry.

Complexity consumes not only budget but also attention. It gives the appearance of activity while slowly dissolving coherence.

Architecture as the Arbiter of Tooling

Rationalization is not austerity; it is restoration. Its purpose is to refine—ensuring that every component strengthens governance rather than dilutes it. Architecture becomes the arbiter of selection, defining the criteria by which technologies are evaluated, integrated, and retained.

Each tool must serve a clearly bounded purpose within a defined control domain such as identity, observability, or data protection. Within each domain, one authoritative system should act as the control plane—the canonical source of enforcement and evidence. Tools that cannot emit verifiable proof of who acted, when, and under what scope, do not belong in a governed environment.

Evaluation begins not with features but with traceability. A capability that cannot produce measurable evidence cannot be governed, and a control that cannot be governed cannot be trusted. Architecture therefore filters tools through intent, ensuring that enthusiasm for novelty does not outpace the system's ability to verify itself.

Rationalization restores architectural hygiene, clarifying what scale has obscured and distinguishing capability from clutter. When conducted regularly and transparently, it aligns investment with principle so that innovation proceeds within the confines of trust rather than at its expense.

Reflect:
When evaluating a new technology, do you ask what it displaces or only what it adds?

Human and Cultural Dimensions

No architecture succeeds through logic alone. Complexity proliferates not only through technology but through behavior. Teams pursue speed in isolation, implementing local solutions faster than governance can respond.

The architect's responsibility is not to suppress autonomy but to shape it, defining boundaries within which independence remains safe. Transparent evaluation criteria and shared metrics for configuration quality, change control, and integration success transform governance from obstruction into collaboration.

Culture determines whether rationalization endures. Where governance is misunderstood, it will be bypassed; where it is understood as a framework for progress, alignment follows naturally. Architecture must therefore communicate intent continuously, ensuring that governance is internalized and not just imposed.

Reflect:
Do your teams understand why rationalization exists or only that it is required?

Architectural Integration and Shared Control Planes

Rationalization reaches its highest expression in unified control planes. Shared systems for identity, network, logging, and policy governance are not administrative conveniences; they are the mechanisms through which governance scales.

When enforcement logic, identity inheritance, and telemetry operate through integrated planes, governance gains both speed and precision. Incidents can be correlated across domains, baselines applied uniformly, and remediation propagated without translation loss. Fragmentation, by contrast, narrows visibility. Each domain reports on its own condition without reflecting the enterprise state as a whole.

Architecture must therefore balance centralization and federation. Autonomy retains value only when it remains measurable. Every system, even those delegated to specific teams, must connect to a common verification fabric so that its activity can be evidenced within the enterprise model of control.

Simplification is not reduction but alignment—the act of ensuring that diversity serves design rather than disorder.

Reflect:
Do your control planes converge on shared evidence, or do they coexist as separate interpretations of current state?

The Architectural Imperative

Complexity becomes inevitable only when architecture abdicates. Rationalization restores clarity by allowing the system to observe itself—to trace intent through execution to evidence. It transforms governance from reactive oversight into active containment, ensuring that each retained component reinforces integrity rather than eroding it.

The integrity of architecture is measured not by scale but by coherence: how consistently components enforce shared logic, how seamlessly telemetry converges into visibility, and how reliably evidence can be reproduced. Every addition, exception, or retirement either strengthens or weakens that coherence.

The decisive question for leadership is not how many tools exist, but how many contribute demonstrable proof of control. Those that cannot should be retired. Complexity unmanaged is not innovation; it is entropy disguised as progress. Architecture's mandate is to make that distinction visible, ensuring that what remains in the system is defensible.

Reflect:
If you had to justify every tool in your environment by its measurable contribution to assurance, how many would remain?

Closing Observation

Complexity is the shadow of innovation. Without boundary and verification, it grows faster than governance can evolve. Rationalization transforms that trajectory from expansion to refinement, proving that architectural clarity—not accumulation—is what sustains progress.

Chapter 5—Multi-Cloud Governance, Baselines, and Boundaries

The modern enterprise no longer resides within a single provider. Through acquisition, autonomy, and the gradual sediment of pragmatic choice, it operates across multiple cloud environments—each with its own identity model, policy dialect, and enforcement logic. What began as technological diversity has matured into structural disparity: a constellation of parallel systems that cannot describe, let alone defend, themselves as one. Under the banner of agility, diversity without design becomes fragmentation.

Architecture's purpose is not to resist that diversity but to civilize it—to translate plurality into coherence through baselines, boundaries, and a unifying geometry of control. Without integration, each platform defines its own standard and risk calculus. Governance ceases to be enterprise-wide and dissolves into local dialects of security, each authoritative only within its own horizon. What disappears first is not control but the consistency that makes control measurable.

Reflect:
How consistent is your organization's definition of "secure configuration" across providers, and are the differences deliberate or the result of drift?

The Architectural Imperative for Consistency

In a multi-cloud ecosystem, consistency is not preference but precondition. Each provider names its environments differently—accounts, subscriptions, projects—but beneath the vocabulary lies the same problem: how to preserve uniform expectation when enforcement logic is distributed across incompatible systems.

A baseline forms the grammar of that language. It defines the non-negotiable controls: consistent encryption standards, comprehensive telemetry, review intervals that keep identity current, and data classifications that preserve lawful boundaries. These are not administrative conveniences; they are the connective tissue through which governance retains shape.

When architecture enforces a unified baseline, posture becomes measurable as a whole. The enterprise begins to reason in terms of total risk rather than provider variance. In its absence, each platform drifts under its own gravity, producing uneven maturity and eroding visibility.

Discrepancies between clouds are evidence of architectural fracture. Consistency, therefore, is not the standardization of technology but of expectation. Only by defining how control should behave everywhere can deviation anywhere be meaningfully understood.

Reflect:
Are your baselines defined consistently across cloud environments, or reinvented for each provider?

Defining and Enforcing Boundaries

If baselines articulate expectation, boundaries embody enforcement. They are the geometry that separates autonomy from exposure, defining where integration begins and ends and under what conditions. In the multi-cloud continuum, boundary discipline distinguishes federation from contagion.

Identity segmentation prevents any single provider from becoming a universal key. Network isolation transforms connectivity from default to privilege. Data jurisdiction encodes geography as policy-as-code. Operational accountability assigns explicit stewardship so that every environment has an owner whose authority and duty are auditable.

When boundaries are implicit, governance reverts to interpretation. When explicit, incidents can be contained, traced, and remediated with certainty. Boundaries are where architecture proves its discipline, converting theoretical control into measurable behavior.

Reflect:
Can you identify the enforcement boundaries for every environment, and who owns verification of their integrity?

Multi-Cloud Policy Integration

Each provider speaks its own syntax of control—AWS Service Control Policies, Azure Policy, Google Organization Policy.

While each can express complex rules, none can interpret the others. Architecture must therefore align intent, not syntax.

Policy-as-Code repositories serve as the translation layer where shared principles are rendered in provider-specific form. Each policy is written in its native language yet derived from a single canonical definition of intent. Architecture ensures that these translations remain faithful, so that enforcement diversity does not become divergence.

Without that alignment, inconsistency becomes inevitable. A rule blocking public storage in one platform but absent in another creates more than variation—it creates uncertainty of assurance. Multi-cloud governance succeeds only when policy intent is defined once, expressed contextually, and verified through telemetry confirming equivalent outcomes across all environments.

This alignment preserves both individuality and integrity: providers retain their strengths while the enterprise retains a unified definition of control.

Reflect:
Is policy intent defined centrally and implemented contextually, or rewritten independently by platform?
How do you verify that equivalent intent produces equivalent outcomes?

Cross-Cloud Telemetry and Evidence

If policy defines behavior, telemetry defines proof. Unified telemetry functions as the nervous system of governance, transforming signals from distinct clouds into a coherent view of health. Logs, configurations, and compliance data must converge into a common observability fabric where context replaces isolation and analysis yields evidence instead of conjecture.

Cross-cloud telemetry strengthens detection, response, and validation. It exposes drift, maps control coverage, measures incident frequency, and tracks remediation time uniformly across providers. Without such unification, each environment becomes its own island, and eventually its own blind spot.

Centralized telemetry converts oversight into assurance. It ensures that control data is not merely collected but correlated, enabling architecture to verify integrity continuously across the multi-cloud fabric.

Reflect:
How frequently does your organization reconcile telemetry to confirm that architectural integrity remains intact?

Architectural Balance: Independence and Integration

Each provider brings distinct capabilities and constraints. The architect's task is not to erase difference but to balance it—achieving coherence without conformity.

Over-standardization breeds stagnation; under-standardization breeds chaos. Effective architecture calibrates both, preserving innovation while maintaining assurance. This equilibrium selectively centralizes what must remain uniform and federates what can safely evolve. Shared controls retain coherence even as each platform advances at its own pace.

The balance is not aesthetic—it is evidentiary. Independence must still produce verifiable proof that intent holds across boundaries.

Reflect:
Can your enterprise demonstrate that policy intent is enforced equally across providers? When one drifts, how rapidly is equilibrium restored?

The Architectural Outcome

Multi-cloud governance matures not when diversity disappears, but when it behaves coherently. Architecture transforms fragmentation into federation through shared baselines, explicit boundaries, unified telemetry, and coordinated enforcement.

Baselines sustain consistency. Boundaries sustain containment. Telemetry sustains visibility. Unified through intent, these disciplines allow the enterprise to operate as one governed organism rather than a mosaic of reconciled reports.

Closing Reflection:

Can your enterprise demonstrate compliance as a single living system or only as a compilation of partial assurances?

Chapter 6—Governance in Practice: Azure, AWS, and the Architecture of Control

A well-architected environment is not one where nothing goes wrong; it is one where what goes wrong is detected and contained. Governance anticipates imperfection and builds structure around it, answering not only *How do I enforce compliance?* but also *What happens when compliance fails?* That answer separates documentation from discipline and bureaucracy from resilience.

Azure and AWS represent two expressions of architectural control. Azure emphasizes centralized posture and identity management with cross-cloud visibility and deep collaboration integration. AWS emphasizes autonomous automation and end-to-end capability management through tightly bound services and intrinsic resilience. Their focus and mechanisms differ, but the architectural objective is the same: enforceable controls that remain observable and recoverable at scale.

Azure Example—Governance Through Integration and Dependency

Azure's governance model reflects a foundation in collaboration and service interconnection. Unlike platforms designed primarily for internal workload execution, Azure extends deeply into the productivity layer—servicing Exchange,

SharePoint, Teams, and other data environments protected by Microsoft Purview. These services blur the line between infrastructure and collaboration, drawing users, guests, and data from multiple organizations into a shared compliance perimeter. Governance in Azure therefore must coordinate not only workloads but relationships: who owns the data, who can access it, and under which identity authority those actions occur.

Microsoft Purview exemplifies this dynamic. It acts as a coordination layer enforcing compliance across Microsoft's broader ecosystem, yet its posture depends on conditions beyond its own control. Default Purview accounts within Azure can expose connectivity from all networks, allowing unintended access unless private endpoints and encryption are explicitly required through **Azure Policy**.

This dependency extends through Azure's wider governance stack. **Entra ID** and **Defender for Cloud** serve as core control engines, delivering centralized identity, posture management, and cross-cloud visibility. **Purview** builds on that foundation to protect services that often reside outside Azure's infrastructure boundary. The result is powerful reach, but also potential inheritance risk. Directory administrators in Entra ID may gain unintended access to Purview or collaboration configurations through cascading privileges. Without strict role segmentation and clear administrative boundaries, directory authority can permeate compliance environments unintentionally.

Conditional Access and related identity safeguards strengthen its model but are frequently applied inconsistently.

Non-human identities such as service principals, automation accounts, and pipelines often operate with persistent privileges that Conditional Access cannot govern. Unless restricted through **Managed Identity** or identity-protection controls, these actors remain invisible to the same safeguards applied to human users.

Defender for Cloud extends Azure's governance outward, detecting configuration drift and posture deviation across Azure, AWS, and GCP. Its visibility is multi-cloud, and many controls can be enforced directly through integrated policies and connectors. Yet full remediation often depends on actions implemented within each platform. Governance remains centralized in oversight but shared in enforcement.

Azure's model succeeds when its dependencies—identity, compliance, and visibility—remain synchronized. It excels at managing hybrid and collaborative ecosystems but inherits fragility from the same integration that gives it strength. Its design is inclusive by intent, bridging users, clouds, and services under a unified compliance narrative.

Reflect:
Are you certain that those with directory authority cannot indirectly modify your collaboration compliance settings?

AWS Example—Governance Built for Autonomous Automation

Where Azure extends governance outward to unify collaboration and cross-cloud visibility, AWS directs governance inward, emphasizing containment, automation, and platform self-sufficiency. Its design centers on native alignment within its own ecosystem, integrating compute, storage, identity, and security controls into a cohesive framework. AWS favors internal integration over hybrid interoperability, optimizing for depth, scale, and operational speed within its perimeter.

This structure accelerates innovation through systemic coherence. Machine learning, analytics, and event-driven automation are core capabilities, supported by encryption that functions as a default architectural requirement. **AWS Key Management Service (KMS)** enables secure cross-region data replication within the platform, supporting expansion without external dependencies in most cases. Secrets management illustrates this layered integration: secrets are stored in **Secrets Manager** with resource-specific policies, encrypted through KMS, accessed via scoped IAM roles, and constrained by organizational policy.

At the organizational layer, **AWS Control Tower** extends layered logic to account governance across the enterprise. **Service Control Policies (SCPs)** can be configured to define global conditions that the individual accounts cannot override, preventing creation of unencrypted resources, restricting long-term IAM credentials, and blocking unauthorized modification of encryption settings. These controls enforce uniformity and containment.

Automation converts these principles into operational behavior through code—repeatable, testable, and enforceable at scale. **Infrastructure-as-Code** (IaC) templates can mandate private endpoints for Secrets Manager, require encryption for sensitive S3 buckets, and integrate compliance validation directly into CI/CD pipelines. In combination with properly configured SCPs, IaC automation ensures deployments begin from a validated, known-good baseline.

AWS telemetry completes the loop. **Config** records the system's declared state; **Security Hub** consolidates evidence; **GuardDuty** and **Detective** perform threat detection and investigation. Together, they transform operational data into demonstrable evidence of governance in motion, converting configuration and event telemetry into verifiable validation of control performance. AWS prioritizes depth over breadth, favoring exhaustive internal observability rather than external aggregation. Its architecture is intentionally introspective, with native services working in concert to sustain continuous verification and containment.

Resilience in AWS arises from this self-containment. Disaster recovery, scalability, and AI-driven remediation are intrinsic capabilities rather than external integrations. This makes AWS highly adaptive within its own environment but in some cases it can seem less fluid across hybrid landscapes, where its closed design can limit interoperability.

When policy, automation, and telemetry remain aligned, AWS governance operates as a coherent internal system of control—efficient, measurable, and structurally self-sustaining.

Reflect:
Do your systems support and enforce sound architecture for business continuity and disaster recovery across regions?

The Architecture of Containment and Resilience

Governance is not the absence of error but the presence of containment. It is the architecture's capacity to remain coherent as the systems beneath it evolve. Azure and AWS may differ in syntax, tooling, and emphasis, but both require resilience engineered into design rather than applied afterward. Holistic enterprise governance reaches maturity when control endures across platforms, automation layers, and organizational boundaries.

The measure of architecture is not how completely it can secure one provider, but how consistently it can sustain assurance across all of them. Azure and AWS illustrate complementary models—one expansive and integrative, the other contained and autonomous—but neither alone defines resilience. In multi-cloud environments, effectiveness depends on whether architecture can translate intent faithfully between them, maintaining consistency even as each enforces control in its own way.

Closing Reflection:

Can your governance maintain consistency and assurance across providers, or only within each one's boundaries?

Chapter 7—Zero Trust and Continuous Verification

Governance, in its most mature form, depends less on the boundaries of individual platforms than on the coherence of its architecture. The same logic that unifies Azure and AWS must also govern the relationships among users, workloads, and data—where connections, transactions, and identity signals provide evidence that assurance is earned continuously rather than assumed.

Zero Trust is not a framework appended to architecture; it is architecture practiced at its most disciplined state. Too often it is reduced to an identity model, when in fact it is an architectural principle that treats trust not as a static attribute but as a conditional state—contextually derived and perpetually verified. Identity remains essential, yet it represents only one facet within a larger geometry of control extending across data, workloads, networks, automation, and human behavior.

Its premise is simple: no entity, human or machine, operates without verification or context. Its practice demands the fusion of policy, telemetry, and automation so that verification becomes the medium through which the system operates. Architecture gives Zero Trust its body, translating the imperative *never trust, always verify* into measurable design. Without that structure, Zero Trust remains aspiration rather than discipline.

Reflect:
How much of your governance model rests on assumption rather than evidence?

Context Over Identity

In a mature system, verification occurs within context. Identity alone is static—it confirms who a subject claims to be, not whether that identity should act here, now, and under these conditions. Context supplies the missing dimensions of judgment: device health, data sensitivity, workload integrity, network path, and behavioral deviation combine to form a composite view of risk that no credential can convey on its own.

Without context, trust decays. A token valid at noon may signal compromise by midnight if reused from an unrecognized endpoint or through unsanctioned sessions. Effective architecture embeds contextual evaluation directly into policy engines, telemetry pipelines, and runtime decision logic so that each access request is assessed continuously and relative to surrounding activity. Trust becomes a dynamic state— earned through evidence, withdrawn the instant that evidence no longer holds.

Reflect:
Which contextual signals inform your access decisions today, and how quickly could your system recognize when a credential presents a red flag?

Continuous Verification as a Control Loop

Verification, properly understood, is not a gate but a loop. Every interaction between users, systems, and workloads must remain observable, evaluated, and adjusted as context changes. In a Zero Trust model, this loop operates continuously: authenticate, authorize, monitor, adapt. Policy engines absorb telemetry, update risk scores, and adjust permissions in near real time.

When the loop functions correctly, posture remains both dynamic and stable—responsive enough to adapt yet consistent enough to contain. When it fails through static credentials, cached sessions, or unmonitored tokens, trust reverts to assumption. Stale trust is latency mistaken for safety. Architecture prevents this stagnation by ensuring that each enforcement point both contributes to and consumes telemetry, transforming isolated controls into coordinated feedback.

Reflect:
After access is granted, how often is it re-verified? Do your detection and response mechanisms operate within one continuous loop?

Architectural Enforcement Points

Zero Trust becomes tangible through enforcement points embedded across every architectural layer.

- **Identity:** Conditional access and just-in-time privilege restrict exposure to what is necessary.

- **Network:** Segmentation and adaptive routing enforce deliberate connectivity instead of default paths.

- **Workload:** Signed artifacts and runtime integrity checks confirm that what executes is what was intended.

- **Data:** Attribute-based access and context-aware encryption protect information according to its classification and use.

Each enforcement point emits telemetry that fuels the next cycle of verification. When integrated, they form a coordinated control system capable of detecting and correcting drift in near real time. When fragmented, the seams between them become the true perimeter—thin, brittle, and easily exploited. Zero Trust typically fails not in principle but in translation: in the gaps where policy stops short of execution. Architecture closes those gaps through shared policy engines, common telemetry models, and unified response logic, ensuring that a signal in one layer can trigger defensive action in another.

Reflect:
Do your enforcement points act from a shared policy engine and telemetry source?

Verification Through Automation and AI

Automation extends Zero Trust beyond human vigilance, correlating signals at a scale that manual review cannot sustain. AI amplifies this reach by identifying privilege escalation, behavioral drift, and anomalous data movement that static rules would miss. When governed correctly, automation shifts defense from reactive containment to anticipatory correction.

Yet automation introduces its own paradox: the temptation to trust the machine. If AI operates without transparency or constraint, it recreates the very flaw Zero Trust was designed to remove—confidence without verification. Architecture resolves this tension by embedding limits and lineage directly into automation systems. Every machine-made decision must be logged, explainable, and reversible within policy boundaries. Authority can be delegated but not surrendered.

Governed automation strengthens assurance; ungoverned, it amplifies risk. The system must learn without assuming infallibility, interpreting signals rather than inventing truth. In that balance, architecture ensures intelligence serves control rather than substituting for it.

Reflect:
What boundaries define your automation's authority? Can every AI-driven enforcement action be traced to its evidence and intent?

When the Verifier Learns

Do these principles still hold when the verifier itself begins to learn? Artificial intelligence now participates directly in the trust mechanism. It identifies anomalies, predicts exploit likelihood, and classifies data at speeds no human could match. Yet each advantage introduces new exposure: the same systems that enforce control can also amplify error. AI simultaneously strengthens and destabilizes Zero Trust for each of the Zero Trust Principles. See examples in the table below.

Zero Trust Principles, AI Advantages, and AI Risks

Zero Trust Principle	AI Advantages	AI Risks
Strong encryption and data protection	Automates key management; detects unencrypted data at scale.	Model training or inference may expose sensitive data or metadata.
Comprehensive asset inventory	Discovers unmanaged or shadow assets through pattern recognition.	Expands attack surface via interconnected AI agents and APIs.
Data classification and configuration monitoring	Uses ML/NLP to classify data and detect configuration drift in real time.	Misclassification or model leakage can expose regulated data.
Least-privilege, adaptive access	Dynamically adjusts permissions based on behavior and context.	Creates new trust boundaries between human and AI-driven decisions.
Access point mapping and monitoring	Detects new endpoints and anomalies from vast telemetry.	Greater interconnectivity enlarges potential blast radius.
Multifactor authentication	Identifies hijacking and behavioral anomalies in real time.	Deepfakes and synthetic identities can bypass MFA.
Network segmentation	Dynamically tunes microsegmentation to contain movement.	Model error or poisoning can propagate misconfigurations.
Secure default configurations	Audits and remediates drift automatically.	Automated correction can introduce systemic error at scale.
Timely patching	Predicts exploit likelihood and prioritizes remediation.	Adversaries use the same models to weaponize exploits faster.
System simplification	Identifies redundancy and dependency sprawl for consolidation.	Overreliance on opaque AI systems reduces explainability and oversight.

These interactions define the frontier of modern assurance. The mistake is assuming AI belongs exclusively to defenders or attackers; in reality, it belongs to both. The velocity of advantage now shifts daily. The question is whether Zero Trust can evolve quickly enough to govern a verifier that learns faster than policy can adapt.

Ultimately, the object of protection is not identity, network, or infrastructure, but rather the data around which all control revolves. When that priority blurs, Zero Trust fragments into isolated control domains instead of operating as a unified context of assurance. Properly implemented, it is not achieved once but maintained continuously—re-evaluated with each generation of automation and verified against every new form of autonomy.

Zero Trust, at its most disciplined, is a living architecture that proves its integrity through adaptation.

Closing Reflection:
If your security posture shifts daily, does your governance architecture detect deviation or simply accommodate it?

Chapter 8—AI Risks and Governance within Zero Trust

Artificial intelligence has moved from the periphery of experimentation to the core of enterprise architecture. It now operates within the bloodstream of modern systems, analyzing telemetry, orchestrating workflows, triggering remediation, and increasingly deciding what qualifies as compliance. Properly governed, it amplifies control; left unbounded, it accelerates error. AI has become the most powerful multiplier of both order and disorder within the Zero Trust model.

Within that model, AI cannot exist outside governance. It must adhere to the same architectural disciplines that constrain human decision-making: explicit boundaries, continuous verification, and demonstrable evidence of control. Without those disciplines, automation drifts into autonomy, and autonomy without oversight becomes risk by design.

Reflect:
Where in your environment does AI operate with authority greater than its accountability? If it made the wrong decision at scale, how quickly could your architecture detect and contain it?

AI as a Core Architectural Element

Architecture makes no exceptions for intelligence, human or artificial. Every model, inference service, and orchestration engine consumes data, performs actions, and produces outcomes that affect compliance, reputation, and posture. Each must therefore operate within the same control fabric that governs all workloads.

The architectural question is not *whether* AI adds value but *under what conditions* it is allowed to act: which data sources it can access, which classifications it can handle, how its outputs are validated, logged, and explained, and what escalation paths apply when it intersects with enforcement policy. Undefined parameters create implicit trust, the most dangerous state of all, where code moves faster than it can be verified.

In a mature Zero Trust architecture, implicit trust cannot exist. Validation is continuous, authority is contextual, and every action must be proven before it is accepted as legitimate.

Reflect:
Which of your AI systems has clearly defined operating boundaries and escalation paths?

The Dual Impact of AI on Zero Trust

AI strengthens verification yet simultaneously expands exposure. It accelerates anomaly detection, predicts configuration drift, and automates enforcement with precision unattainable by human processes. Yet the same capabilities that improve assurance can amplify risk. AI can aggregate sensitive data across domains, generate untraceable outputs, or replicate misconfigurations at scale.

Architecture must convert that acceleration into advantage without allowing automation to exceed oversight. Each data connection requires defined boundaries; each model must operate with transparency and produce evidence of its decisions. Each automated action must remain auditable and reversible. The core disciplines of Zero Trust—least privilege, explicit verification, and continuous assurance—form the structural controls that keep intelligence accountable. Properly applied, they allow automation to scale securely rather than autonomously.

Reflect:
Does your architecture measure both the benefits and the liabilities of AI? How is trust for your intelligent systems defined, granted, and revoked?

Architectural Controls for AI Governance

Governance of AI cannot rely on declaration. It must be embodied in enforceable architecture, where every model functions as a governed workload with defined identity, telemetry, and accountability. Architecture must therefore encode the following continuous disciplines:

- **Identity and Ownership:** Each model, dataset, and automation pipeline must have a designated owner responsible for compliance, retraining, and evidentiary traceability.

- **Data Integrity and Lineage:** Training and inference data must remain classified, validated, and traceable.

Lineage ensures that regulated inputs are not reused without review and that data changes automatically trigger reassessment of performance and bias.

- **Operational Boundaries:** Every AI process must operate within policy-defined constraints. Policy-as-Code enforces those boundaries, preventing unauthorized escalation or unapproved data access.

- **Transparency and Explainability:** Decision logic must remain observable and auditable. Opaque models cannot coexist with evidence-based assurance.

- **Lifecycle Assurance:** Each model must include defined criteria for monitoring, retraining, and decommissioning. Bias detection, drift analysis, and performance validation must run as continuous control functions.

When these controls operate together, AI becomes a governed component of enterprise architecture. When they do not, compliance becomes narrative and accountability dissolves into assumption.

Reflect:
Are your AI systems version-controlled, audited, and monitored with the same rigor as your code? Could you demonstrate model lineage and accountability before a regulator or board?

AI as a Verification Amplifier

When governed properly, AI strengthens verification by expanding the scope and velocity of assurance. It analyzes telemetry to detect configuration drift, privilege escalation, and control failure before incidents occur. It consolidates evidence across platforms, transforming compliance from periodic review into continuous validation.

Architecture must, however, preserve control over execution. AI may recommend actions but cannot implement them without independent verification. No inference engine should modify production configurations, adjust policies, or assign privileges without passing through the same approval logic required for human change. Automation enhances defense only when its outputs are subject to the same rigor required to create them.

Reflect:
In your architecture, where does AI advise, and where does it act? What mechanisms verify those actions before they reach live systems?

Policy Drift and Versioned Validation

Even with safeguards in place, automation can drift from its intended design. The same intelligence that verifies compliance can, under new data or shifting context, begin to redefine it.

Policy drift, once the product of human neglect, now arises from automated misalignment. A misclassified dataset, an exposed API, or a recursive feedback loop may shift enforcement away from architectural intent without any explicit, approved change request.

The containment mechanism for this risk is **versioned validation**. Every control must carry a reference to the model or automation logic that applied or last modified it. This provenance forms a traceable chain of trust through code, configuration, and inference.

Automation must not outpace governance. Systems that patch, reconfigure, or revoke privileges automatically must also record *why* those actions occurred and *under which logic* they were justified. Explainability and rollback capability must be embedded in every layer of infrastructure- and policy-as-code. An automated action without a human-readable rationale is indistinguishable from a compromise. Every AI-enabled control must operate within defined governance boundaries. Every function must correlate directly to an authorized requirement. Examples:

AI-Enabled Controls and Governance Requirements

Control Mechanism	AI-Enabled Function	Governance Requirement
Anomaly Detection	Model flags unusual access patterns	Validate thresholds; require dual approval before enforcement.
Remediation Automation	AI triggers configuration rollback	Record rationale and retain pre-change state.
Access Recommendation	Adaptive engine suggests privileges	Treat as advisory until verified by policy tools.
Data Classification	NLP-driven tagging of data	Sample for accuracy; block propagation without review.

Zero Trust, viewed through this lens, focuses less on enforcing a static standard and more on maintaining alignment among autonomous systems. The principle of *no implicit trust* remains, but the object of verification expands. Governance must validate not only *who* acts but *how* the system interprets policy.

Identity frameworks must now include AI entities, complete with credentials, scopes, and revocation controls. Data governance must treat models and embeddings as sensitive assets. Policy engines must record model provenance and confirm that automation operates under validated logic sets.

Reflect:
If a model trained six months ago continues to enforce policy, is it enforcing your current standards or a version that no longer applies?

Adaptive Trust Assurance

Observability must also expand beyond system telemetry to machine reasoning itself: data inputs, inference metadata, decision logic, and error propagation. This is not merely auditability; it is containment through comprehension. A model whose actions cannot be explained is equivalent to an unverified identity.

Architecture becomes a negotiation between autonomy and control. Every model and orchestration pipeline introduces a new node of delegated judgment. Each must be measured against a unified trust baseline. AI systems may generate insight faster than humans, but if not bound by explicit policy logic, intelligent control devolves into intelligent exception.

Zero Trust in this environment cannot merely assume breach; it must assume adaptation. Verification evolves into **continuous behavioral correlation**, monitoring not only for compliance but for divergence from intent. Drift detection, lineage validation, and rollback automation together define the next discipline of governance: **Adaptive Trust Assurance**—a model in which trust measures its own alignment dynamically across human and machine actors.

The architecture that endures treats intelligence, human or artificial, as a governed resource. Zero Trust does not dissolve under automation; it scales through it. What remains constant is explicit structure: defined boundaries, validated logic, and reversible action. The future of governance depends not on trusting AI to secure the system but on ensuring the system remains secure even when AI is wrong.

When Intelligence Becomes a Principal

Artificial intelligence did not create new principles of security; it exposed the limits of the old ones. For decades, trust architecture relied on predictability—authenticated identities, segmented networks, encrypted data. But a system that learns is a system that changes, and governance must now treat learning entities not as tools but as *participants* in the trust equation.

AI introduces the **machine principal**: entities that are not users yet authenticate; not administrators yet provision, classify, and remediate at scale. Verifying them requires more than credentials—it requires lineage, logic visibility, and intent validation. Every model functions as a form of identity; every inference a potential transaction. These identities demand full parity with human ones: credentials issued, scopes defined, privileges revocable.

Each layer of architecture inherits new dependencies as intelligence embeds deeper. Mappings at the architectural layer define where governance must expand to include machine actors, examples seen in the table below.

Architectural Layers, AI Dependencies, and Zero Trust Reinforcements

Architectural Layer	AI Dependency Introduced	Zero Trust Reinforcement Required
Identity and Access	Service principals, model roles, API tokens	Enforce role isolation; require signed inference requests; audit model privileges.
Data and Storage	Training data, embeddings, feature stores	Track lineage; restrict input sources; encrypt model artifacts.
Network and API	Model endpoints, vector databases	Segment inference boundaries; require private endpoints and tokenized transfer.
Automation and Orchestration	Auto-remediation scripts, ML actions	Enforce rollback logic; require approval for policy propagation.
Visibility and Analytics	Model observability, telemetry fusion	Correlate behavior with baselines; detect divergence between human and machine policy interpretation.

Traditional enforcement was transactional: authenticate, authorize, log. AI-era enforcement is continuous and probabilistic. The system must verify not only *who* acts but *how* that action was derived. Telemetry must capture decision states as well as resource states, allowing architecture to audit reasoning with the same rigor as results.

Containment, in this context, ensures integrity—the assurance that learning systems remain within visible bounds of governance. It prevents unauthorized propagation of logic, unverified decisions, and invisible breaches. Without containment, AI evolves faster than the frameworks that govern it, becoming too opaque to audit and too autonomous to correct. Architecture introduces deliberate friction where evidence lags behind action, slowing acceleration just enough to keep proof aligned with outcome.

Reflect:
How visible are your AI systems' decision flows and data exchanges? Could you revoke an AI's operational authority without collapsing dependent processes?

Zero Trust as the Anchor of AI Governance

Zero Trust supplies the control framework within which AI governance operates. It defines the conditions for trust—how intelligence authenticates, how authority is constrained, and how actions remain verifiable. Within this structure, no AI process should operate without authentication, authorization, and auditability. Every AI-driven action must be reviewable and reversible; every dataset and inference must comply with classification and encryption requirements; every automation must act within visible, bounded policy.

AI is already embedded in the control fabric, analyzing telemetry, managing privilege, and triggering remediation— often ahead of the safeguards that should govern it. Within

Zero Trust, its function must evolve from automation for efficiency to automation for verified assurance. Architecture's role is not to decide *whether* AI participates but to ensure that participation remains transparent, accountable, and reversible. Without those boundaries, intelligence accelerates risk faster than governance can contain it.

Closing Reflection:
How many AI systems in your environment operate beyond explicit verification or visibility? Could you prove today what data they were trained on, and whether those sources still meet your compliance standards?

Chapter 9—Responsible AI, Data Lifecycle, and Ethical Governance

Artificial intelligence has moved from experimentation to infrastructure. Often it no longer merely supplements enterprise systems; it defines how data flows, how controls respond, and how governance is applied. As intelligence becomes operational, architecture must evolve not only to secure what systems do but to ensure that the reasons they act remain visible, explainable, and correctable. Governance now extends beyond managing systems to stewarding intent.

Responsible AI is not a declaration but a structural property of architecture. It measures integrity through evidence—the verifiable connection between data origin, model behavior, and decision outcome. Designing responsibly means embedding transparency and accountability into every operational layer through controls that validate themselves, telemetry that records provenance, and automation that remains bound to the principles that authorize it.

Reflect:
Does your AI governance framework produce measurable evidence or only statements of intent?

The Architectural Responsibility of Data Stewardship

Every intelligent act begins as a data event. The fidelity of that event—the accuracy, bias, and context of the information captured—determines both the reliability and the fairness of the decisions derived from it. Architecture defines the conduits through which data flows, the classifications that govern its use, and the lifecycles that constrain its persistence. Data quality determines not only accuracy but justice; without that foundation, no ethical control can stand.

A single misclassified dataset can corrode years of maturity, and unlabeled records can embed privacy violations that no algorithmic refinement can later erase. In disciplined environments, data provenance and lineage are not metadata but control points embedded within the same logic that enforces access and encryption. Architecture demands structural answers: Where did this data originate? Who authorized its use? Under what consent, retention, and destruction rules does it exist?

Embedding those answers into operational systems through policy-as-code, metadata registries, and automated cataloging transforms governance from awareness into enforcement.

Reflect:
Can your enterprise trace every dataset used for AI back to its lawful source, its risk classification, and its consent record?

Ethical Governance as a Structural Control

Ethics in AI is an operational parameter that converts human principle into system constraint, ensuring that automation remains subject to deliberate authorization. It addresses not only whether a model *can* make a decision but whether it *should*—and under what authority, with what oversight, and through which escalation path.

The answers manifest architecturally: API gateways that block unreviewed inferences; workflows that require human concurrence when outputs affect welfare, legality, or reputation; and policies that establish transparency thresholds for contestable decisions. Such mechanisms express governance as verifiable infrastructure, translating intent into enforceable design.

Reflect:
Are ethical constraints encoded in your architecture or merely discussed in your policy documents?

The Data Lifecycle as an Architectural System

Responsible AI requires that data exist within a closed, auditable loop. Architecture defines this lifecycle as a chain of accountability:

1. **Acquisition and Classification:** Identify each source, classify by sensitivity and purpose, and bind it to consent metadata.

2. **Usage and Retention:** Authorize access only to approved models within explicit time windows; record every utilization event as evidence.

3. **Transformation and Training:** Apply minimization, anonymization, and bias screening before ingestion to ensure that learning reflects legitimate context.

4. **Archival and Disposal:** Impose retention limits and irreversible deletion once purpose expires, preventing informational persistence beyond necessity.

Each phase generates telemetry: lineage records, validation proofs, and cryptographic attestations that feed continuous compliance analytics. A disciplined lifecycle ensures that data does not persist beyond the context that justified its creation and enforces the principle of *finite trust*.

Reflect:
Can your architecture demonstrate when, how, and why every AI dataset enters and exits the system?

Bias, Explainability, and Human Oversight

Bias cannot be corrected through principle alone; it must be detected, measured, and constrained through architecture embedded within the same telemetry that governs compliance and drift. When evaluation is continuous—testing parity across population segments, sources, and time—organizations can distinguish structural imbalance from incidental error and correct it before it calcifies into automation.

Assurance depends not only on what a system decides but on how it reached that decision. Each inference must disclose the data that informed it, the features that influenced it, and the thresholds that triggered its outcome so governance can test reproducibility, trace bias, and validate compliance claims. Version control, feature attribution, and inference logging together form the evidentiary chain through which architecture sustains assurance and enables credible challenge when outcomes are contested.

Yet even in functional and auditable systems, human interpretation remains essential. Data may describe; it cannot judge. Architecture must define where human authority re-enters as a condition of legitimacy, ensuring that algorithmic precision does not replace ethical evaluation.

Reflect:
Where does human judgment intersect your automated decision flow, and how is that oversight enforced through architecture rather than assumed through policy?

The Convergence of Ethical and Operational Assurance

Ethical governance and Zero Trust share a single foundation: trust is not declared or assumed but verified. Responsible architecture applies that principle equally to conduct and control, requiring evidence not only that systems function but that they do so within lawful and ethical bounds.

Telemetry must therefore extend beyond operational metrics to include indicators of fairness, consent, and value alignment. When these signals are measured through the same instrumentation that tracks security and performance, ethical assurance becomes an operational discipline rather than an aspiration.

Reflect:
Can your architecture demonstrate ethical compliance with the same clarity it demonstrates system reliability?

The Architectural Mandate for Responsible AI

Ultimately, Responsible AI is an act of architecture—an insistence that intelligence remain governed by structure and bound by the same requirements that secure identity and network segmentation: explicit boundaries, measurable enforcement, and continuous validation. Its distinction lies in scope, spanning the technical, legal, and ethical domains simultaneously.

Enterprises that confine AI governance to documentation will see their assurances erode at the speed of automation. Those that embed it in design will find that ethics becomes durability—the visible proof of integrity in every system decision. Responsible AI does not constrain innovation; it stabilizes it, ensuring that trust remains demonstrable and renewable through design.

Closing Reflection:
Can your organization prove, with evidence, that every

intelligent system operates within ethical, legal, and architectural boundaries?

Chapter 10—Continuity, Assurance, and Culture as Control

Architecture survives not through permanence but through renewal. Its coherence depends on whether discipline can outlast the people who defined it. Continuity, then, is not the archiving of design but the transmission of intent through a system that can verify itself even as context shifts and interpreters change. Assurance provides the proof of that continuity, demonstrating that what was built still behaves as intended under new conditions. Yet neither continuity nor assurance can persist without culture—the operational conscience that keeps verification a practiced discipline rather than a procedural formality.

A mature architecture endures not because it resists change but because it absorbs it without losing integrity. It replaces dependence on memory and individual expertise with patterns, automation, and evidence strong enough to survive rotation. Continuity, assurance, and culture are not abstractions; they are the architecture's immune system, preserving coherence as everything around it evolves.

Reflect:
Would your architecture survive if the people who designed it moved on?

Continuity as a Property of Design

Continuity is not a passive condition but an architectural capability. It exists when systems can reconstruct themselves from first principles, verify their integrity, and correct drift. Architecture achieves this when its controls, policies, and dependencies are encoded as dynamic components—versioned, observable, and testable against a defined standard.

Where that encoding is absent, erosion begins quickly. Configurations diverge, controls weaken, and institutional knowledge dissipates as teams or platforms change. Each transition compounds the loss, forcing organizations to rediscover what they already solved. This is not simply a human lapse but a structural one: a design that assumed continuity instead of engineering it.

Continuity must therefore be deliberate, achieved through automation that embodies principle in reproducible form:

- **Infrastructure-as-Code** rebuilds environments consistently, achieving recovery through execution.

- **Policy-as-Code** carries standards forward as personnel and priorities shift.

- **Continuous-compliance pipelines** transform detection from periodic audit to active self-verification.

Through these mechanisms, governance becomes regenerative—a system that sustains alignment with intent regardless of who operates it or how often it changes.

Reflect:

Could your environment be rebuilt identically and securely without relying on the memory of a few key people?

Assurance Beyond Audit

Traditional assurance captures compliance at a point in time, verifying that controls once aligned with policy. Architecture, however, operates continuously, and its validation must do the same. Assurance cannot depend on retrospective checks; it must confirm, in real time, that controls function as designed and that deviations are detected and corrected before they degrade posture.

Telemetry provides the foundation for that verification. Control data flows through validation pipelines that measure both presence and performance. Deviations trigger remediation rather than reports. Metrics such as drift rate, control coverage, and mean time to correct indicate the operational health of governance itself. When verification is embedded directly into operation, assurance becomes an architectural property rather than an administrative exercise.

Continuity depends on evidence; assurance depends on verification; culture ensures both remain habits.

Reflect:

Does your assurance model measure control performance as it happens, or only record that it once met standard?

Embedding Governance into Operating Culture

Technology enforces discipline; culture preserves it. Without cultural reinforcement, even the best architecture fragments under change. Governance must therefore function not as compliance oversight but as instinct—a shared understanding that security, configuration, and automation express the same structural integrity.

A mature operating culture demonstrates three conditions:

1. **Accountability is systemic.** Ownership of configuration and automation is explicit. Boundaries define who may deploy, approve, or override. Engineers recognize that automation extends authority and must therefore be governed as rigorously as identity.

2. **Verification is continuous.** Drift detection, deviation review, and control validation are treated as operational metrics, not audit artifacts. Feedback loops measure governance as naturally as performance.

3. **Learning is corrective.** Every incident, exception, and AI misjudgment informs architecture, refining boundaries so that error cannot recur under automation.

Without these conditions, automation scales misconfiguration as efficiently as control. AI-driven systems begin to act without verification; policies diverge; boundaries blur until configuration becomes indistinguishable from compromise. Culture, more than technology, prevents self-governing systems from becoming ungoverned. It sustains the discipline that automation enforces and ensures that execution remains subordinate to the authority that defines it.

Reflect:
Does your operating culture reinforce governance boundaries and verification loops, or rely on tools to compensate for absent discipline?

Automation and the Human Element of Assurance

Automation preserves governance through consistency and speed. It enforces policy and verifies outcomes at a scale no human could match. Yet without understanding, it also accelerates error.

Machines can enforce rules; only people can interpret truth. Algorithms perceive data as it exists, not as it is skewed, incomplete, or unjust. Human oversight determines where context ends and consequence begins. Architecture must therefore embed deliberate points of human re-entry—not to repeat automation's labor but to ensure that execution remains aligned with intent, consequence, and ethical constraint.

Assurance falters when the human layer disengages, when automated results are accepted without interpretation or when review becomes procedural rather than analytical. Sound architecture ties accountability to visibility: each responsible actor must be able to see, measure, and explain the risks within their domain. Governance remains machine-enforced but human-directed, precision informed by judgment.

Reflect:
Do your teams interrogate assurance data to understand risk, or merely record it to satisfy compliance?

Architecture as an Enduring Practice

Continuity is the defining measure of architectural maturity. Standards and controls hold meaning only when they outlast the people and circumstances that created them. Once institutionalized, architecture ceases to be documentation and becomes reflex—a system that preserves coherence through structure and disciplined adaptation.

A mature architecture translates discipline into resilience. It persists through disruption because its logic is encoded, its evidence self-verifying, and its governance distributed across people, process, and code. This is not rigidity but confidence— the ability to evolve without losing integrity.

Assurance, then, is not faith in technology but confidence in the discipline that keeps technology honest. Architecture becomes the enterprise's immune system: containing risk before it propagates, learning from each event, and reinforcing the mechanisms that sustain trust.

Closing Reflection:
If every person who built your architecture left tomorrow, would the system still prove its own integrity?

Epilogue—Evidence of Discipline

Discipline is the architecture's foundation; evidence is its result. What begins as structure must end as proof.

Architecture, at its highest expression, is not a diagram of technology but a system of discipline—a structure through which an enterprise continuously proves what it believes to be true about itself. Trust is not granted but earned through evidence: repeatable, transparent, and independent of memory or proximity.

Every preceding discipline—governance, rationalization, Zero Trust, responsible AI, and assurance—converges on a single question: **Can the system prove that it still behaves as designed?** That question separates confidence from control and aspiration from stewardship.

Leadership's responsibility is to own the integrity of systems—to ensure that control, automation, and dependency can be defended in evidence. That duty demands transparency even when uncomfortable, standardization even when inconvenient, and the retirement of tools that no longer contribute to measurable assurance.

The enduring responsibility of leadership is to sustain an environment that produces truth faster than risk can obscure it. Innovation, speed, and scale only matter when matched by a proportional capacity for verification. Architecture must remain a living instrument of evidence: adaptive, measurable, and accountable.

Closing Reflection:
If evidence is the language of integrity, can your architecture speak for itself?

The measure of architecture is its capacity to sustain integrity through change—to analyze, constrain, and reveal the systems it governs faster than risk can erode them.

Printed by Libri Plureos GmbH in Hamburg,
Germany